/'Nosos - 'Lethe/

"Diseased – Oblivion"

James J Tate

/'Nosos - 'Lethe/

"Diseased – Oblivion"

"Save the Animals, Save the World!"

James J Tate

James J Tate

ISBN: 978-1-7366097-0-5 (Book)

For information contact:

www.diseasedoblivion.com

https://youtube.com/channel/UCH_Hei3CbKT7gV88ctd1hUA

jamesjtate7@gmail.com

10 9 8 7 6 5 4 3 2 1

Table of Contents

James J Tate

Dedications

To my family and friends.

Also, especially to Animals.

James J Tate

Preface

This book came to fruition because I had a realization, is this actually what I wanted to do with my life? Am I pursuing the purpose in life that I wanted or actually attended to be?

I wrote *"**Nosos-Lethe**"* because of the unjust conditions that animals have been having to live in. I am a witness am compelled to speak out, I could no longer just keep silent. I wanted to put together a persuasive argument that would change the tides of impossibility into reality.

I was motivated by putting together something good, and it naturally came at ease being inspired to speak out for such loving animals.

Why am I the one to deliver this message? I have always had this connection with animals, for some reason, they have given me an instant acceptance and adored me. Not to sound strange but maybe they knew all along that I was going to write this out and deliver this most important message?

I hope my readers will be impacted by my writing and that the content is taken to enrich lives.

Nosos-Lethe

I wanted to create a word that would describe a specific condition, of an ongoing circumstance that would capture the essence of its nature. I am compelled to express these ideas and perplexed on why discussions of this situation are not being addressed, as it should be?

The title ***"Nosos-Lethe"*** means "Diseased-Oblivion", it is the best description I could put together that is going to give meaning to what is expressed in the following text. To conceptualize a complete full-scale understanding, I am going to add an emotional value "Humanities Greatest Shame!"

Let me began at an axiom that sets forth a pinnacle of conditions that defiled Human morality. Since Humankind utilized Animals for fur or food, a sacrifice paid from Animal suffrage is accrued in blood by these innocent creatures. Accumulating all the Animal suffrage and death intentionally caused by Humanity, and continues is in its current state, as a Systematic Institutionalized Capitalist market of suffrage and death specifically of Agricultural Animals such as Cows, Chickens, and Pigs. Then combine all the deaths caused by heart disease with humanity from these actions; I will say "the oblivious indifference is a disease of thought, as it is the diseased-oblivion for both Humankind and Animal kind". I am grief-stricken and overcome with sorrow thinking about all the lives born to experience this kind of existence. Humankind utilizes Animals in this existence, Humankind no longer needs the suffrage or death of Animals to exist. Your existence is not more important than these living Animals!

Humanity

This is cause-and-effect Humanity, and our place on this planet is being challenged by undesirable controllable actions that can be corrected. Humanity is accountable; being passive, disregarding, or neglecting to understand the importance of this topic, your abandonment and resistance to change is endangering you and everyone else. How many more deadly pandemics must we make it through, is there a chance we face one that will make us extinct? If someone denies that the pandemic, man-made or not but derived or mutated from existing viruses is not a manifestation of this current topic, then they most likely will deny Global Warming too. The history of pandemics keeps repeating itself, humanity is condemned to this fate if things do not change, wake up!

So, if you want to rationalize or justify, just for clarity to be very precise, let me scream in your head "you are committing bloody murder, you are the face of oppression!" If for you, the whole unpleasantness of taking a life is removed, and the physical act is not done by your hands, the guilty exploitation of Ag Animals' is still a depraved act equal to the action, and you are still accountable for it.

What is the relationship that Humanity has with Ag Animals? I believe that the natural state of mutualism where both species benefit from the relationship, only by living together would be a healthy goal. What is the existing symbiotic relationship now, is it parasitism and predation? I have heard the argument that we are biologically designed omnivores as justification for Ag Animal meat-eating. If it is our natural place as omnivores maybe a civilized approach to the plate might be called for, some diplomacy. Invite a bull to dinner and with your bare hands and teeth have a bit of some rump, not seeing that conversation going over so well for you. Better yet, just because you can does not mean that you should. Being that we are omnivores, that means that there is a choice; you can choose to eat Ag Animal meat or the preferable choice of not eating Ag Animal meat. If you do, why would you choose to eat Ag Animal meat? Do you truly want these actions to reflect the reality, you believe your character to be, isn't this moral choice weighing on your conscience? I asked myself the same questions, my choice was very heavily based on morality, but is my appetite or what goes across my taste buds more important? I obviously choose not to eat Ag Animal meat and glad to remove toxins from my diet. I hope most of Humanity would like to live in a world free of the extortion of Ag Animals' suffrage and death.

I still get triggers; I have had my fair share of setbacks but every day just like anything else, I make the efforts to control my diet. Imagine replacing any of your best meals or your favorite items that are Ag Animal-based for just as tasty plant-based items, it is happening! A great meal is going to satiate your satisfaction, appetite, and desire, they are out there!

I will use this similar situation as an analogy, when a call for change happened, smoking was socially acceptable, then everyone became aware of the hazards and the hazardous environment of smoking is prohibited in most public places. Plant-based alternative food is not just about removing hazards, it is about replacing it with something much healthier. Everyone's experience is going to be different, might not even feel the change, you are not going to die, improvements will happen over time. If addicts can make it through addictions this can be possible too. Many reasons why you should go through with all this trouble, like on my cover "Save the Animals, Save the World". Let me spotlight a different point of view, the reciprocity of emotional currency, Humanity benefits so much from Animals for such a long time in so many ways, there is a debt Humanity owes and needs to be paid back to Animals, use that as a driving force.

I had this interesting argument with a friend, he went on to express his opinion that humane acts are from one Human to another explicitly. I went on to explain to him that to be humane are attributes of Human character. A few descriptions of being humane are sympathetic, compassionate, kind, giving, and caring, those are inert dispositions. This world is going to be better off when Humanity is empathic enough to truly live up to our character.

"How Now Brown Cow?"

My interview with a Brown Cow.

James "Do you believe you are happy cows?"

Cow "Come on, what the f#%k do you think James?"

James "Tongues are wagging, people have some pretty bad things to say about you cows". "There is diminutive talk saying you are nothing but Animals, and that you are just fat Cows".

Cow "Let me grab the bull by the horns on this one James". "Some people think that our only value is only being Cash Cows, and they are going to milk it for everything we've got". "Let me tell you how the Cow ate the cabbage". "They can save their prejudice, bias assessed judgment; their discrimination has manifested itself to be this, as their solution". "Let me play devil's advocate, if Humans were being put through the meat grinders how acceptable would that be?"

Cow "We sacrifice the meat off our bones, we give the skin off our backs, when is enough, enough?" "We have families and heard life full of meaningful purpose, we have earned the right to a peaceful existence".

Cow "Not to milk the bull, but people are so bullheaded, we wouldn't want to cow anyone into doing the right thing when the milk is free right?"

Cow "Not to be cynical but don't let me get started on all the people who build their Ag Animal lives around us and their opportunistic bullsh!t".

James "So don't have a Cow, stop eating Ag Animal meats, you have heard it straight from the horse's mouth or should I say the Cow's mouth".

The Hunter

Who are these Hunters, can we put a mirror to these individuals? Wouldn't their primary purpose have been to provide food for survival, and given value to them by achieving life-sustaining skills for themselves and others? It appears to have been a statured position that over time has become a tradition and an elite sport coveted by Royalty, prestigious Aristocrats, and Land Owners? But we all know an image is not everything, and in fact, appealing to tradition because it has been that way in the past is a fallacy.

In the harshest of conditions, the wild animals have the rigor to survive, the competition for Human survival in the wilderness is over.

There is no need to pull the trigger, choose life over death, being capable of such a high level of violence is not a practice that should be exercised by any Human being. These actions are premeditated murders, societies have deemed it one of the worst actions of human behavior. What kind of person seeks the thrill of the hunt, what is the real reason someone would like to get their hands bloody, are any of these conditions' favorable motives? It would be easy to ridicule some of these individuals as being ***psychopaths*** but these are real burdens, and they should have real consequences to live with. Not like the individual who is remorseful because they have been caught, but actual grief for the harm that they have caused and to put down their arms'!

Agency Urgent Accountability!

I have briefly viewed some of the U.S. government agencies webpage's like the United States Department of Health and Human Services, the Food and Drug Administration, the Center for Disease Control and Prevention, and the United States Department of Agriculture; the most relevant and occurring theme is that in their mission statements, protecting public health is their main concern. From what I have seen is that heart disease is the leading cause of death in America. From what I have gathered is that if there is a reduction of one single ingredient of saturated fats, the immediate removal of Ag Animal meats, which is loaded with saturated fats from the American Diet, could it possibly save numerous amounts of American lives? So, no matter how much wealth and power these private or corporate companies have attained, isn't there some petition that can be established to a law, that will prohibit their death-dealing? To make it illegal to kill Ag Animals whose meat is distributed, that is directly linked to heart disease, the leading cause of deaths in America? We have cut out the use of lead in paint even though kids loved to eat it. I would say let us cut off the lard, and this should go for any community anywhere, for the immediate dismantlement of the Animal Ag Industry.

Closing

I understand that the things that I have said are controversial, and that is the whole point, to open a discussion. We as a people need to put a spotlight on this topic.

I believe in Humanity, and we will put an end to Animal cruelty. If you feel that the things that I have expressed are inflammatory or upsetting please share your thoughts with me, maybe there is a point of view or perspective, I am not seeing or understanding? I also believe that the truth and justice will be the guiding moral compass to end this horror!

Book Cover

When I designed the front cover, I wanted to make it into something that was art and symbolic to my cause. I started with some beautiful animals and made them into (round reliefs) symbols or as logos for my work. With the title ***"Nosos-Lethe"*** I wanted it to have the look of a word defined out of a dictionary, and because it was derived from Ancient Greek words, I put in checkers like a lot of the tile work that they did to tie it in as a tribute. The Arch of Constantine is a representation of the past and what is to be of the future. As for the future, I put the sun and rays as a guiding light of the truth that Ag Animals should not be murdered or killed.

Made in United States
Troutdale, OR
11/03/2026

56845111R00021